NISTIR 7506

Impact of a Residential Sprinkler on the Heat Release Rate of a Christmas Tree Fire

Daniel Madrzykowski

Fire Research Division
Building and Fire Research Laboratory
National Institute of Standards and Technology
Gaithersburg, MD 20899-8661

May 2008

Department of Homeland Security
Michael Chertoff, Secretary
Federal Emergency Management Agency
R. David Paulison, Administrator
United States Fire Administration
Gregory B. Cade, Assistant Administrator

U.S. Department of Commerce
Carlos M. Gutierrez, Secretary
National Institute of Standards and Technology
James M. Turner, Deputy Director

ABSTRACT

Although the number of Christmas tree fires is low, these fires carry a higher level of hazard than other fires that occur in a residential structure. This study, supported by the U. S. Fire Administration, has the following three objectives: 1) characterize the heat release rate of dry Fraser fir trees 2) demonstrate the ignition resistance of a tree with a high moisture content and 3) examine the impact of a residential sprinkler on the heat release rate of a dry tree that is on fire in a compartment.

The heat release rates of the trees which were allowed to dry ranged from 3.2 MW to 4.3 MW. Trees that were kept in water, so that the needles maintained a moisture content in excess of 100 %, self-extinguished after being exposed to a flaming book of matches.

The data from the furnished sprinklered room experiment demonstrated that even under conditions of extreme fire growth, a single sprinkler was able to prevent flashover and limit the spread of fire to other objects. The peak heat release rate, from the sprinklered room, was limited to approximately 1.8 MW. The furnished non-sprinklered room experiment generated a post-flashover heat release rate in excess of 6 MW.

Keywords: Christmas trees, fire prevention, heat release rate, heat flux, mass loss rate, residential sprinklers

DISCLAIMER

Certain companies and commercial products are identified in this paper in order to specify adequately the source of information or of equipment used. Such identification does not imply endorsement or recommendation by the National Institute of Standards and Technology, nor does it imply that this source or equipment is the best available for the purpose.

TABLE OF CONTENTS

	Page
ABSTRACT	iii
INTRODUCTION	7
EXPERIMENTAL APPROACH	8
Tree Conditioning	8
Moisture Content Measurement	9
HEAT RELEASE RATE EXPERIMENTS	10
Maintained Trees	11
Dry Trees	11
Results	12
ROOM EXPERIMENTS	16
Results	19
DISCUSSION	25
SUMMARY	27
ACKNOWLEDGEMENTS	27
REFERENCES	27

INTRODUCTION

The National Fire Protection Association (NFPA) reports that there is an estimated annual average of 210 home structure fires that begin with Christmas trees. Based on data from 2002 through 2005, these fires caused an average of 24 civilian deaths, 27 civilian injuries, and $13.3 million in direct property damage per year [1].

The NFPA analysis also shows that although the number of Christmas tree fires is low, these fires represent a higher level of hazard. On average, 1 of every 9 Christmas tree fires resulted in a fatality compared to an average of one death per 75 non-confined home structure fires overall. Further, 49 % of Christmas tree fires spread beyond the room of origin. The fires that spread beyond the room of origin caused 94 % of the associated fatalities [1].

The percentage of trees involved in structure fires represent an extremely small portion of the total number of natural Christmas trees sold, which is estimated at 30 million trees, in the United States each year [2]. The moisture content of each tree can play a dominant role in determining the fire hazard each tree represents. Research by Chastagner on evergreen twigs and by White et. al. on complete trees has shown that fresh cut branches or trees, that have been properly watered and maintained, i.e. trees (needles) with a high moisture content, are difficult to ignite with a small flame [2-4].

The moisture content of a fresh cut tree is typically in excess of 100 %. Once harvested, the tree's rate of drying is dependent on the species of the tree, the initial condition of the tree, and the environmental conditions that the tree is exposed to after being cut. Environmental conditions include temperature, vapor pressure deficit and wind, during transport, retail display and use of the tree [2]. Guidance for caring for a cut tree is provided by the National Christmas Tree Association [5].

Previously, the National Institute of Standards and Technology (NIST) conducted a series of fire tests to characterize the potential hazard from ignition of dry Scotch Pine Christmas trees. Heat release rate was measured as a function of time from ignition using an oxygen consumption calorimeter. Seven trees were allowed to dry for approximately three weeks. These trees burned easily when ignited with a small open flame. Peak heat release rates ranged from about 1.6 MW to 5.2 MW [6]. These heat release rates for dry trees were significantly higher than heat release rates for "Christmas" trees that had been published prior to that time [4,7].

An eighth tree, kept in water until just before testing, could not be ignited. Prior to testing, the moisture content in each tree was measured at several points on the trunk using a moisture meter. The moisture content of the trunk ranged from 25 % to 35 %. While the moisture content of the tree trunks varied by only 10 %, the first seven trees differed significantly in apparent dryness of the needles and branches from the eighth tree.

The 1999 NIST Scotch Pine Christmas tree report also offered some qualitative guidance on determining if the tree is dry and prone to an easy ignition based on the condition of the needles. The needles on the first seven trees were dry to the touch and slightly brown in color. The needles broke easily when bent slightly and fell from the tree when the branches were shaken. The eighth tree had a much greener color and the needles were very pliable. When a needle was pulled, the tree branch would bend slightly before the needle would separate [6].

EXPERIMENTAL APPROACH

This series of Fraser fir experiments were conducted with the following three objectives. The first two objectives extend from the 1999 NIST study: 1) characterize the heat release rate of dry Fraser fir trees and 2} examine the ignition resistance of a tree with a high moisture content. The third objective was to examine the impact of a residential sprinkler on a dry tree that is ignited within a furnished compartment.

Tree Conditioning

Twelve Fraser fir trees were purchased from a commercial tree lot in Mt. Airy, MD on November 27, 2006. The trees had been harvested from a tree farm near Gettysburg, PA on November 17, 2006. Half of the trees were placed in a tree stand with no water. The other six trees were prepared for display as recommended by the National Christmas Tree Association [5]. The preparation includes cutting approximately a 13 mm (0.5 in) disk of wood from the base of the tree trunk. The cut is made perpendicular to the axis of tree trunk, as recommended. After the tree trunks were cut, the trees were placed in reservoir type tree stands. The reservoir has a capacity of 4.42 L (1.17 gal). The trees were kept in the NIST Large Fire Laboratory in Gaithersburg, MD. Water was added daily as needed to keep the reservoirs full.

The weather in the Gaithersburg area was mild during the time the cut trees were being conditioned. While the trees were kept inside the fire laboratory, the laboratory has limited heat and humidity control. During hours of operation, the laboratory environment tends to follow the outside conditions as significant amounts of make-up air are required when the exhaust hoods are operating. Weather data from November 2006 through January 2007 is provided in Table 1 to provide a sense of the conditions.

As demonstrated by previous studies, the fire safety of a tree is directly related to moisture content. [2-4,6]. Chastagner has shown that Fraser fir branches will ignite and burn when exposed to an open flame at a moisture content of approximately 30 % or less [3]. The trees that were not maintained with water were showing signs of moisture loss in mid- January, after being relocated to an area in the building adjacent to a gas-fired space heater for a week. The signs of moisture loss included needles that would easily fall off of the tree, some discolored needles, and ends of branches that were becoming brittle and would snap instead of bend.

Table 1. Weather Conditions for Gaithersburg, MD, November 2006 – January 2007 [8]

	November 2006	December 2006	January 2007
High Temperature (°C/°F)	23.4 / 74.1	23.4 / 74.1	20.3 / 68.5
Low Temperature (°C/°F)	-3.3 / 26.1	-7.5 / 18.5	-10.0 / 14.0
Average Temperature (°C/°F)	9.4 / 48.8	5.6 / 42.0	3.8 / 38.8
High Pressure (hPa / in)	1035.8 / 30.59	1035.8 / 30.59	1040.5 / 30.73
Low Pressure (hPa / in)	995.8 / 29.41	997.9 / 29.47	1002.6 / 29.61
Average Relative Humidity (%)	67	57	56

Moisture Content Measurement

On the day the trees were fire tested, the moisture content (MC) was measured using an Arizona Instruments Computrac Moisture Analyzer, Model MAX 200XL. The apparatus measures the mass of needles removed from the tree. Then the needles are heated to a temperature just above 100 °C until a steady mass reading is obtained indicating that the moisture has been removed.

The MC is determined by the difference of the initial (wet) mass minus the dry mass. The difference is then divided by the dry mass. The resulting value is multiplied by 100 to yield the percent moisture content as shown in equation 1 below.

$$MC = \frac{Mass_{wet} - Mass_{dry}}{Mass_{dry}} \times 100 \qquad [1]$$

Using the moisture analyzer apparatus, the dry trees had needle MCs in the range of 6 % to 9 %. The trees that had been kept in a water filled tree stands had MCs of 120 % to 132 % on the day that they were fire tested. The trees that had been maintained with water still exhibited good needle color and retention, and flexible branch tips. The uncertainty of the moisture content of the needles measurement is estimated to be ± 2% [9].

Another measure of MC was made on each tree by using an electrical resistance moisture meter on the trunk sections of the tree. The moisture content of the tree trunks were in the range of 20 % to 35 %. There was no significant difference between the dry trees and the trees that had been supplied with water. The resolution of the meter is 0.1 % MC. The estimated expanded uncertainty is ± 2 % [10].

HEAT RELEASE RATE EXPERIMENTS

The heat release rate experiments were conducted in the NIST Large Fire Laboratory utilizing the 6 m by 6 m oxygen depletion calorimeter. The estimated uncertainty is ± 11% on the measured heat release rate with an expanded uncertainty of 2. Details on the operation and uncertainty in measurements associated with the oxygen depletion calorimeter can be found in [11]. The data was recorded at intervals of 1 second on a computer based data acquisition system.

A single Schmidt-Boelter total heat flux gauge was installed 1 m above the base of the load cell and 1 m away from the edge of the widest portion of the tree. Results from an international study on total heat flux gauge calibration and response demonstrate that the uncertainty of a Schmidt-Boelter gauge is ± 8 % [12].

The tree in a stand was positioned on a load cell with a resolution of a 0.001 kg. The estimated total expanded uncertainty is ± 0.005 kg [13]. The experimental arrangement for the heat release rate experiments is shown in schematic form Figure 1 and in a photograph in Figure 2.

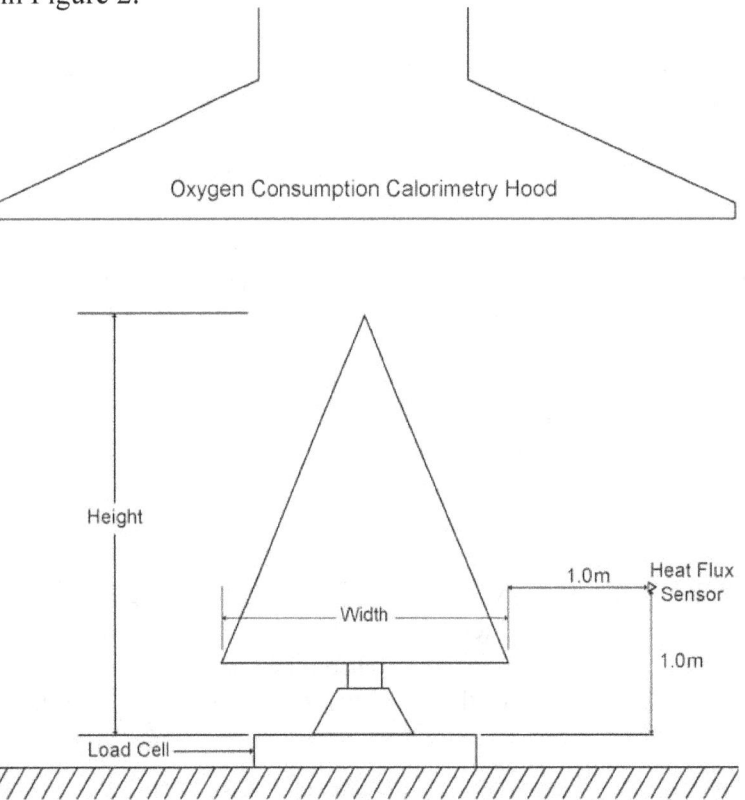

Figure 1. Heat release rate experimental arrangement

Maintained Trees

Ignition experiments were conducted on three of the trees that had been kept in water. The ignition source for the trees consisted of a cardboard book of 20 paper matches that was ignited by an electrically heated wire. The match book was positioned in each tree approximately 0.61 m (2 ft) above the floor and approximately 0.30 m (1 ft) into the tree.

In each test, the flame from the burning matchbook ignited needles in close proximity (within 50 mm or 2 in) of the matchbook, typically a branch with needles that was exposed to direct flame. The fire, however, did not spread and the flames self-extinguished within 90 seconds or less. Each of these trees had MC measures of greater than 100 %. No measurable changes in heat release rate, heat flux, or mass loss were observed during the ignition trials on the high moisture content trees.

Figure 2. Tree in stand on load cell under oxygen consumption calorimetry hood

Dry Trees

Similar ignitions were used on five dry trees. In each case, the matchbook ignition was of sufficient energy to ignite the tree, additional branches became involved, and the tree continued to burn. The dimensions and initial mass of the trees are given in Table 2. The masses indicated for trees 1 through 4 include the mass of the plastic tree stand, 1.20 kg.

Table 2. Dry Tree Measurements prior to testing

Tree No.	Height (m / ft)	Width (m / ft)	Mass (kg / lbs)
1	2.05 / 6.75	1.52 / 5.00	12.37 / 28.31
2	2.54 / 8.33	1.55 / 5.10	13.81 / 30.38
3	2.39 / 7.83	1.52 / 5.00	13.48 / 29.66
4	2.23 / 7.30	1.63 / 5.35	10.97 / 24.13
5	2.06 / 6.77	1.45 / 4.75	14.83 / 32.63

The first four trees were tested in a vertical orientation, as shown in figure 2. The last tree, tree 5, was removed from the tree stand and positioned on the load cell as shown in figure 3.

Figure 3. Tree 5 prior to ignition

Results

Figure 4a through 4d are a series of photographs that show the stages of fire growth from just after ignition to just before total burn out. The span of time from Figure 4a to 4d is approximately 60 seconds. Note the amount of vertical flame spread versus the horizontal flame spread in photograph 4b. In the last photograph of the series, the majority of the tree trunk and many of the larger branches are intact after the needles have burned away.

a. 1 second after ignition b. 10 seconds after ignition

c. 35 seconds after ignition d. 61 seconds after ignition

Figure 4 a – d. Photographs of Tree No. 3

Figure 5 displays the heat release rate time history of each of the five trees. The peak heat release rates ranged from 3.2 MW to 4.3 MW for the trees 1 through 4. The majority of the needles were burned in less than 60 seconds. Tree 5 which was positioned horizontally had a peak heat release rate of 1.6 MW and a longer burn time.

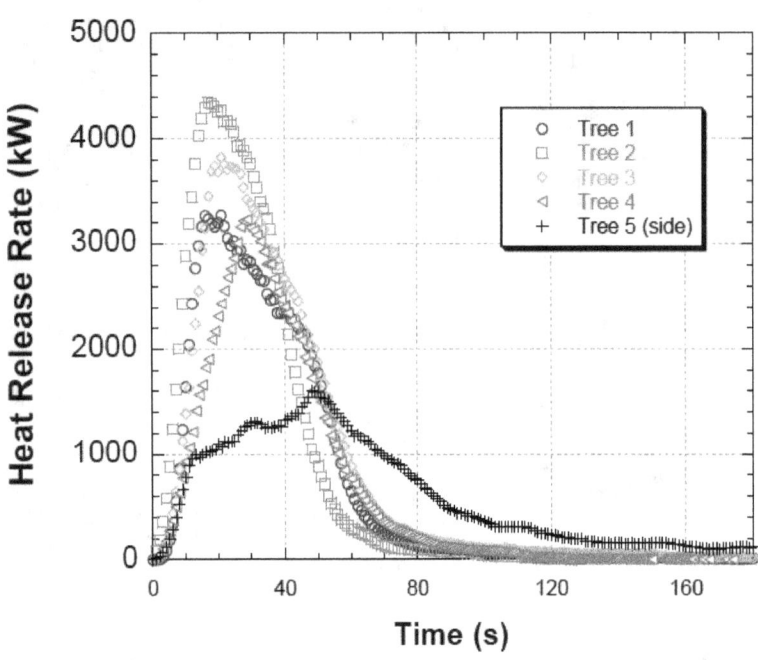

Figure 5. Heat Release Rate vs Time for dry Fraser firs

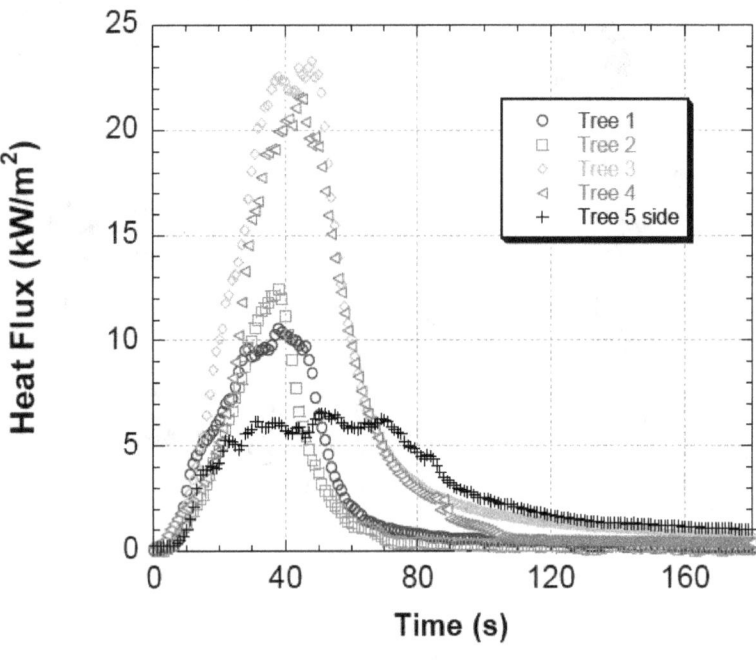

Figure 6. Heat flux vs time for dry Fraser firs

The total heat flux measured from each of the burning trees is shown in Figure 6. The peak heat fluxes for trees 1 through 4 show a variation of a factor of two between trees 1 and 2 and trees 3 and 4. Note that the heat flux peaks occurred after the HRR has peaked. The trees were ignited on the side opposite the heat flux sensor. As a result, the unburned tree branches blocked some of the radiant energy as the fire was spreading up through the tree and the peak heat fluxes occurred as the branches closest to the heat flux sensor were burning. Hence the variation in heat flux is a factor of position and view factor for the given position of the heat flux sensor. The variation is not representative of overall heat flux emission of the trees. In future experiments, additional data can be collected by positioning several sensors around the tree and comparing the averages for each fire.

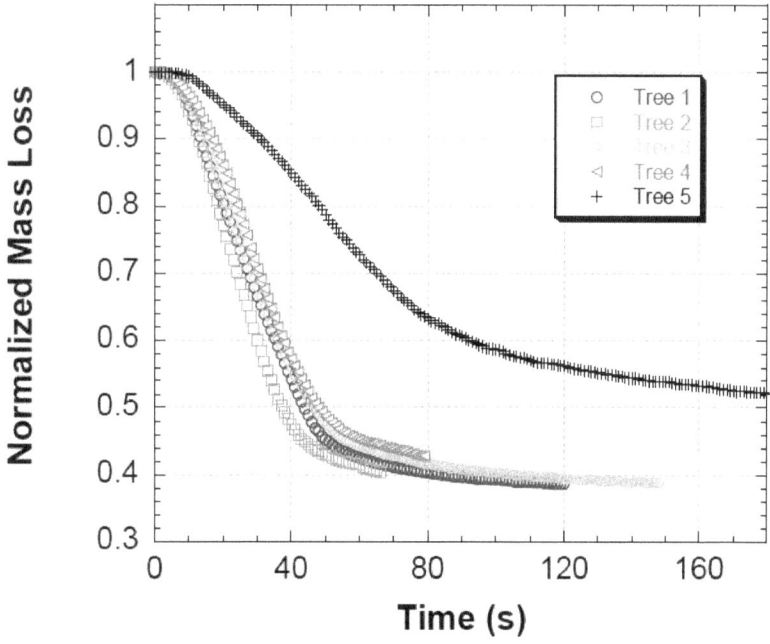

Figure 7. Normalized mass loss vs time for dry Fraser fir trees

Figure 7 shows the normalized mass loss of each tree. The normalized mass loss is the mass of each tree as it is burning divided by the its original mass. The first four trees lost between 55 % and 60 % of their initial mass within 60 seconds after ignition. Tree 5 did not burn as fast or as thoroughly due to its horizontal positioning. During the experiment, branches that were near the base end of the horizontally positioned tree did not burn. This resulted in less mass loss compared to the trees that were in a vertical position.

ROOM EXPERIMENTS

In order to examine the impact of a residential sprinkler on a dry tree, two full-scale experiments were conducted. The first experiment was conducted in a non-sprinklered room. The second experiment was conducted in a room with a residential sprinkler. The sprinkler was installed in accordance with the manufacturer's listing instructions [17].

Both rooms had similar furnishings (size, fabric, etc.) and interior finish. The ceiling and walls were made from two layers of 12.7 mm (0.5 in) thick gypsum board. The exposed surface of the inner layer of gypsum board was painted with two coats of latex paint. The floor of each compartment was composed of 19 mm (0.75 in) thick plywood. The plywood was covered with 12.7 mm (0.5 in) thick polyurethane foam padding. Nylon cut pile carpeting with a polyolefin backing was installed on top of the padding. General material descriptions, overall dimensions and the mass of each item is provided in Table 3. Size and mass of the dry trees used in the experiments is given in Table 4.

Table 3. Room Furnishings

Item	Materials	Dimensions (m / ft)	Mass (kg/lbs)
Padding	Polyurethane Foam	3.63 x 3.45 x 0.012 / 11.9 x 11.3 x 0.04	20.4 / 44.9
Carpeting	Nylon w/Polypropylene backing	3.63 x 3.45 x 0.012 / 11.9 x 11.3 x 0.04	22.4 / 49.3
Sofa	Polyester Fabric, Polyurethane foam seat cushions, Polyester batting filled back cushions	2.09 x 0.86 x 0.75 H / 6.86 x 2.82 x 2.46 H	47.7 / 105.0
Chair	Polyester Fabric, Polyurethane foam seat cushions, Polyester batting filled back cushions	0.81 x 0.86 x 0.75 H / 2.66 x 2.82 x 2.46 H	29.1 / 64.0
Table	Wood	0.61 x 0.71 x 0.60 H / 2.00 x 2.33 x 1.97 H	9.6 / 21.1
Lamp Shade (top dia x bottom dia x H)	Polystyrene	0.18 x 0.48 x 0.31 H/ 0.58 x 1.58 x 1.00 H	0.38/ 0.84
Lamp	Ceramic w/ metal	0.15 dia x 0.55 H / 0.5 dia x 1.8 H	1.13 / 2.5
Framed picture	Wood frame, glass	0.85 x 0.70 x 0.024 / 2.8 x 2.3 x 0.08	5.0 / 11.0
Total			135.7 / 298.6

Table 4. Dry Tree Measurements prior to room experiments

Room	Height (m / ft)	Width (m / ft)	Mass (kg / lbs)
Non-sprinklered	2.40 / 7.87	1.50 / 5.00	14.84 / 32.65
Sprinklered	2.40 / 7.87	1.50 / 5.00	14.64 / 32.21

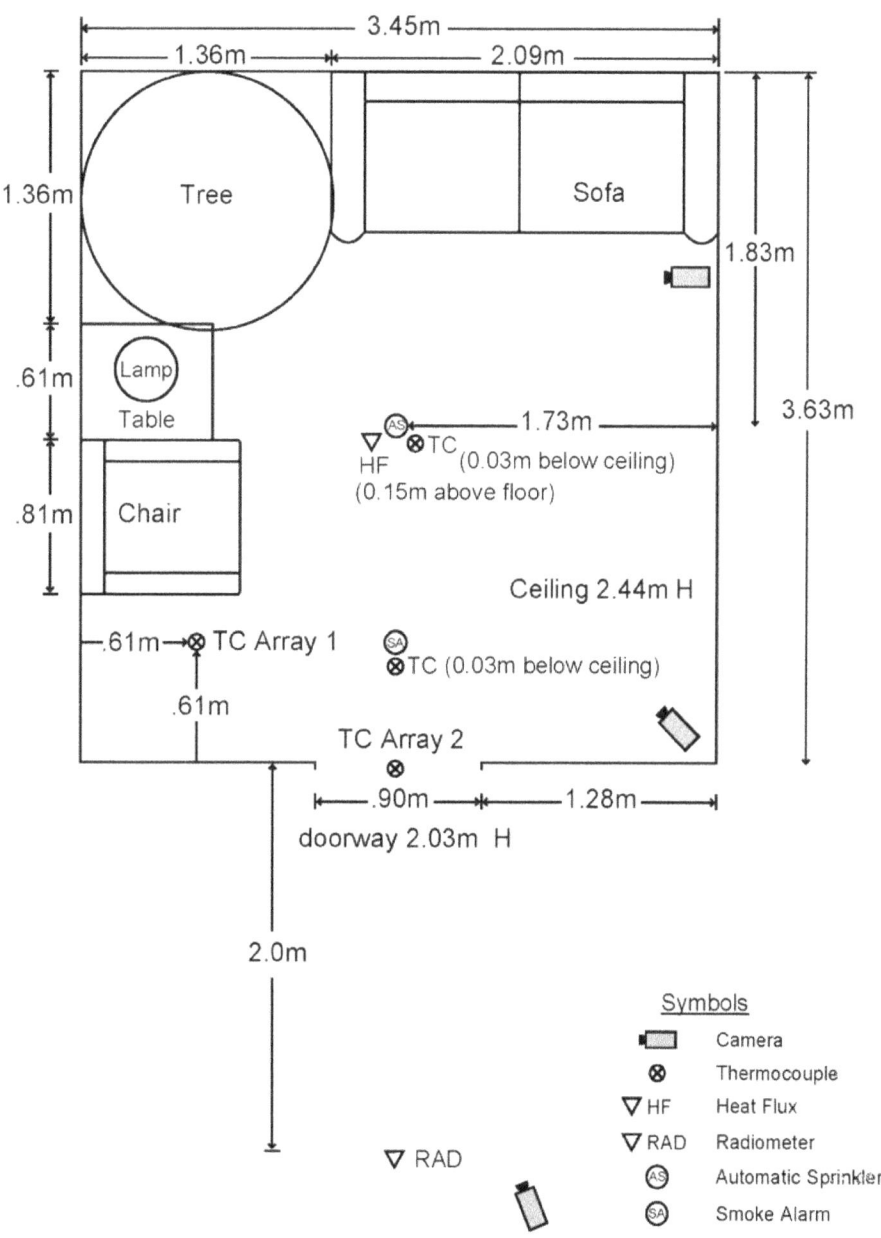

Figure 8. Plan view of furnishing and instrumentation arrangement

Figure 8 shows the dimensions of the test compartment, the arrangement of the tree and furnishings, and the location of the instrumentation. Each compartment experiment was instrumented with thermocouples, a heat flux gage, a radiometer, and video cameras.

Each thermocouple array was composed of Type K, 0.25 mm bare bead thermocouples. One thermocouple array (TC Array 1) was located 0.61 m (2.0 ft) out of the front left corner of the compartment. The vertical array had thermocouples located 0.025, 0.3, 0.61, 0.91, 1.22, 1.52, 1.83, 2.13 m below the ceiling (BC). Another vertical array of thermocouples (TC Array 2) was located in the open doorway. This array had thermocouples located at 0.025, 0.3, 0.61, 0.91, 1.22, 1.52, and 1.83 m below the door lintel (BL).

Single thermocouples were installed 25 mm below the ceiling and adjacent to the ionization smoke alarm and automatic fire sprinkler.

The standard uncertainty in temperature of the thermocouple wire itself is $\pm$ 2.2 °C at 277 °C and increases to $\pm$ 9.5 °C at 871 °C as determined by the wire manufacturer [14]. The uncertainty of the temperature in the environment surrounding the thermocouple is known to be much greater than that of the wire [15, 16]. Small diameter thermocouples were used to limit the impact of radiative heating and cooling.

A Schmidt Boelter total heat flux gauge was installed near the center of the compartment and 0.15 m above the floor. The sensing surface was facing the ceiling. In addition, a Schmidt Boelter type radiometer with a sapphire window and 150 ° view angle was installed 2 m back from the open doorway and 1 m above the floor of the compartment on the outside of the compartment. The sensing surface was facing the open doorway. As noted before, the uncertainty of a Schmidt-Boelter gauge is $\pm$ 8% [12].

The experimental compartments were arranged so that the open doorway was under the 6 m by 6 m oxygen consumption calorimeter. Given that the combustion products collect and mix in the compartment prior to flowing into the oxygen consumption calorimetry hood for measurement, the heat release rate will not provide a precise time history of the fire growth rate. However, the heat release rate data does provide a comparison between the two compartment fire experiments.

Ignition was conducted in a similar manner to the heat release rate experiments. The moisture content of the needles was between 6 % and 9 % for each of the trees used in these experiments. The moisture content of the trunk of each tree was between 25 % and 30 %. These conditions were similar to those of the dry trees used in the heat release rate measurements.

Results

The heat release rate, temperatures from the thermocouple arrays and the total heat flux near the floor of the room for the non-sprinklered experiment are presented in figures 9 through 12 respectively. In the non-sprinklered experiment, the smoke alarm activated 11 seconds after ignition, the sofa was burning approximately 45 seconds after ignition and the compartment transitioned to flashover at approximately 60 seconds after ignition. Manual suppression using a small hose line from outside the compartment was initiated at 90 seconds after ignition.

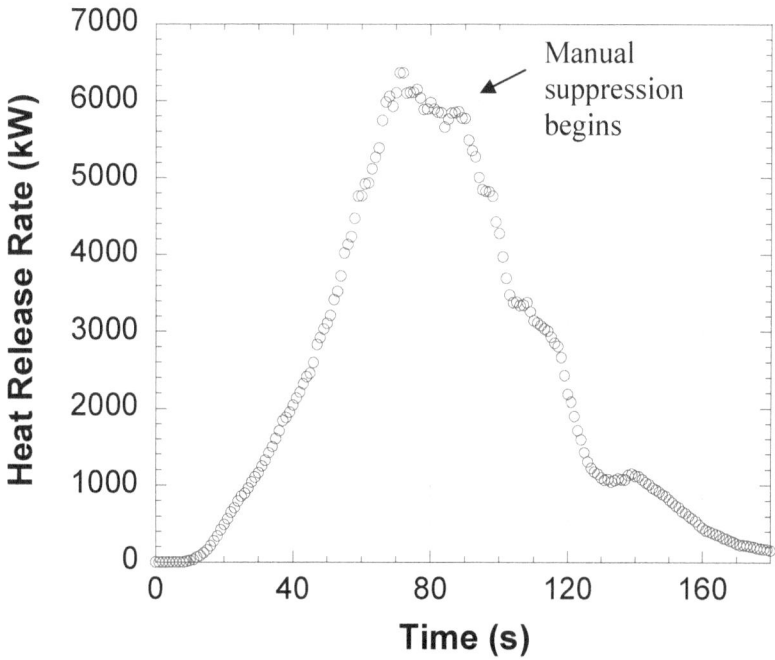

Figure 9. Heat release rate vs. time for the non-sprinklered compartment

The radiant heat flux, measured 2 m from the doorway and 1 m above the compartment floor, outside of the compartment is shown in Figure 13. The peak heat flux of 27 kW/m^2 occurs at approximately 70 seconds after ignition.

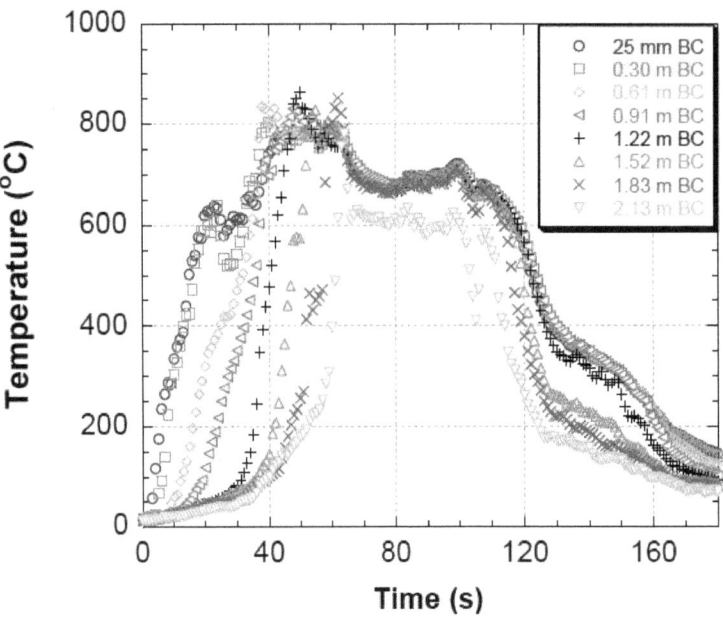

Figure 10. Temperature vs Time, TC Array 1 in the non-sprinklered compartment

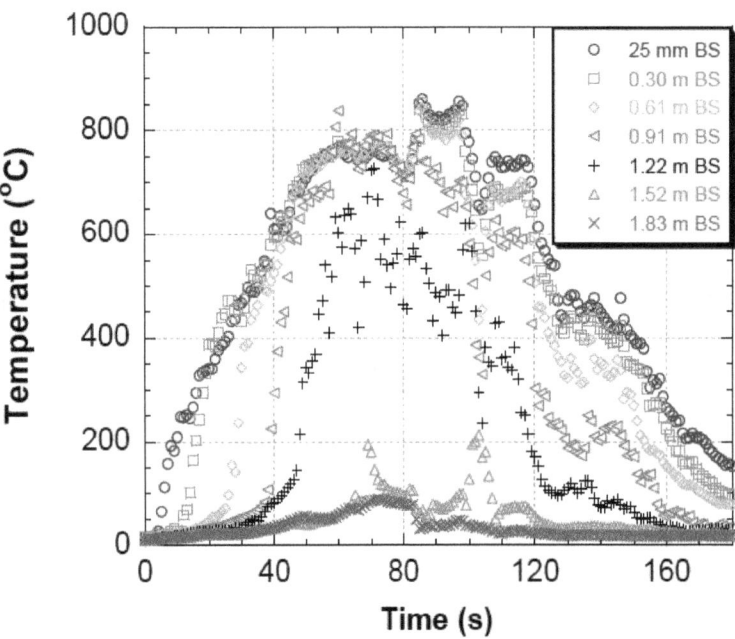

Figure 11. Temperature vs. Time, TC Array 2 in the non-sprinklered compartment

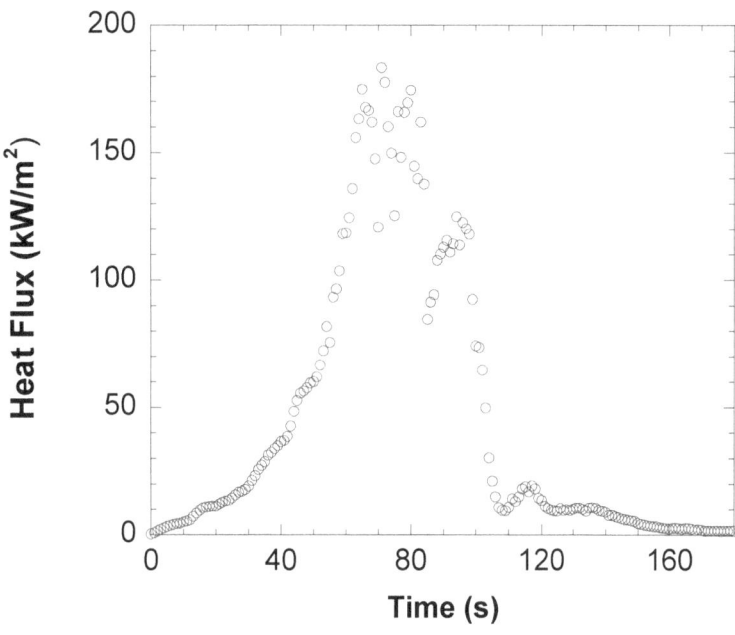

Figure 12. Total heat flux vs time in the non-sprinklered compartment

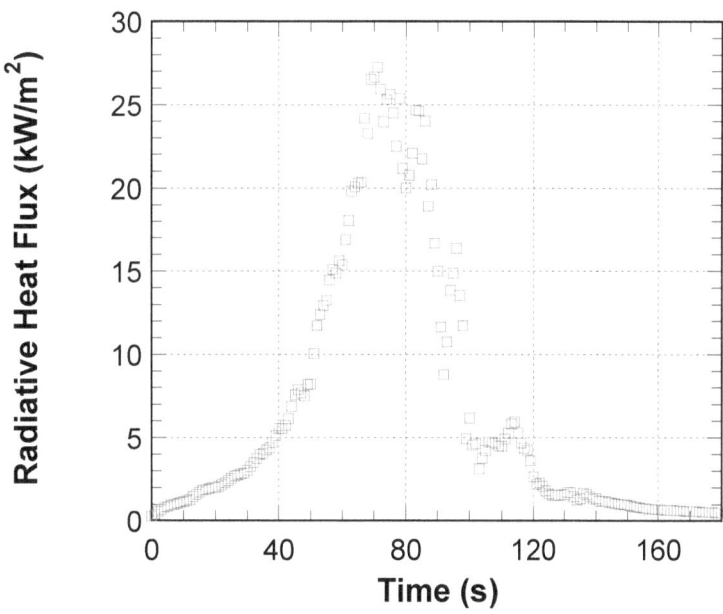

Figure 13. Radiant heat flux outside of the non-sprinklered compartment

The sprinklered compartment experiment had a similar fuel load, dry tree and instrumentation as the non-sprinklered compartment. The significant difference was the installation of a pendent, residential sprinkler. The sprinkler had an activation temperature of 68 °C (155 °F). For a maximum coverage area of 3.7 m (12 ft) by 3.7 m (12 ft), the listed minimum water requirement is 34.1 L/min @ 58.1 kPa (9 gpm @ 8.4 psi) [17]. The water supply to the sprinkler was set to 34.1 L/min (9 gpm).

The smoke alarm and the sprinkler both activated 10 seconds after ignition. The fire developed so rapidly that the temperature of the thermocouple next to the sprinkler measured approximately 500 °C (932 °F) just prior to activation. The thermocouple adjacent to the smoke alarm measured approximately 350 °C (662 °F) just prior to alarm.

The heat release rate and temperatures are shown in figures 14 through 16. The rate of heat release was reduced by the activation of the sprinkler, however, the heat release rate continued to increase as the tree continued to burn. That being said, the water from the sprinkler kept portions of the tree below the point of ignition from burning and limited the flame spread on the furnishings. As a result, the compartment did not flashover and the peak heat release rate was kept to less than one third of the non-sprinklered compartment experiment.

The peak temperatures in the upper half of the compartment ranged from approximately 200 °C (400 °F) to 700 °C (1300 °F). The temperatures, figure 15, in the lower half of the compartment range from 30 °C (86 °F) near the floor to 67 °C (152 °C) at 1.52 m (5 ft) below the ceiling. The temperatures in the doorway are significantly lower than in the compartment as shown in figure 16. Measured heat flux increased less than 0.5 kW/m^2 during the sprinklered experiment.

Photographs in figure 17 show fire development in the doorways of the sprinklered room and the non-sprinklered room. The photographs were taken near the peak heat release rate of each experiment, at approximately 60 seconds after ignition.

Post experiment photographs are presented in Figure 18. In the photograph from the sprinklered room, thermal damage is limited to the tree and wall and ceiling areas adjacent to the tree. Most of the furnishings in the room did not ignite. In fact the lamp was still operating after the fire experiment. A small portion of the sofa in contact with the tree ignited. The fire spread to the sofa was limited to that initial area and was extinguished by the sprinkler within 3 minutes and 30 seconds of ignition. The photograph from the non-sprinklered case shows damage that is consistent with post-flashover conditions, burn damage throughout the room, floor to ceiling. All of the furnishings in the non-sprinklered room ignited.

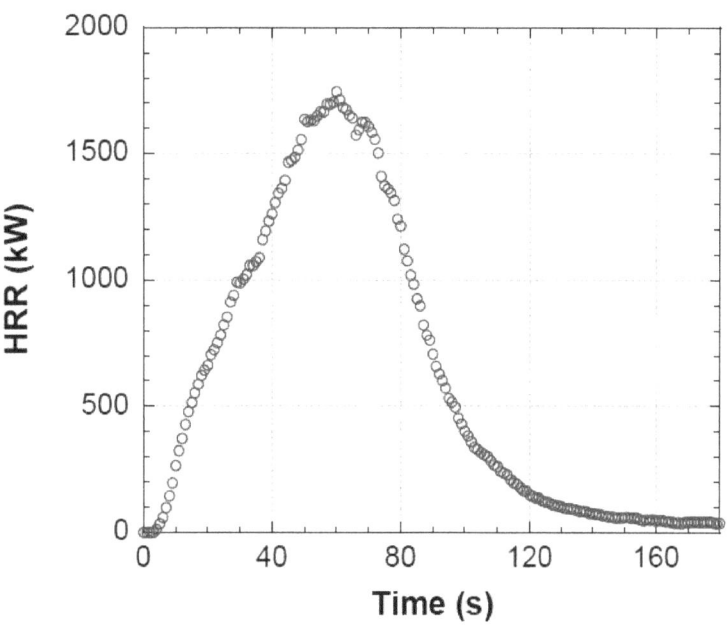

Figure 14. Heat release rate vs time for the sprinklered compartment

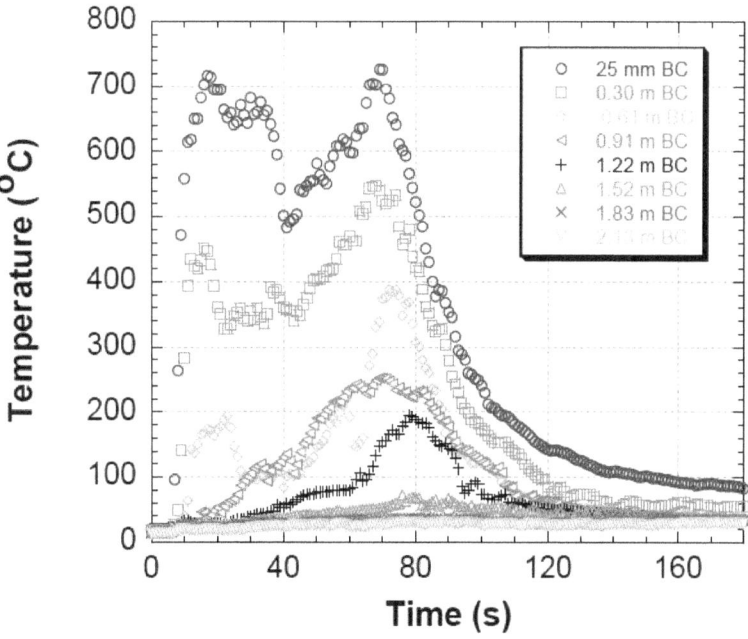

Figure 15. Temperature vs time, TC Array 1 in the sprinklered compartment

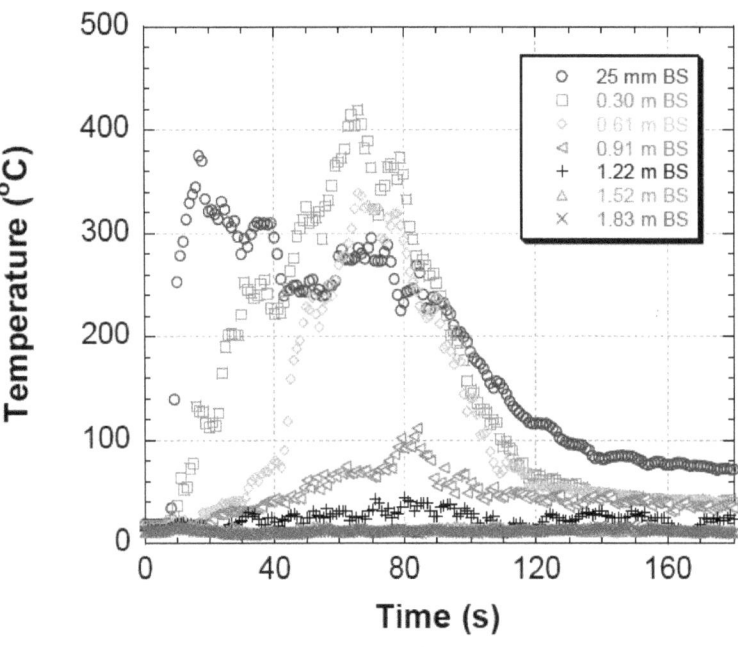

Figure 16. Temperature vs time, TC Array 2 in the sprinklered compartment

Figure 17. Comparison of the sprinklered room, on the left, with the unsprinklered room
(Both photos were taken at approximately 1 minute after ignition.)

Figure 18. Post fire photographs (Sprinklered room on the left and non-sprinklered room on the right.)

DISCUSSION

The results from the Fraser fir tree fires were consistent with the previous NIST study on Scotch pine trees. In both cases, trees with dry needles were readily ignited with a flaming source and generated high heat release rates. The Scotch pines, which ranged from 9.5 kg to 20 kg, generated peak heat release rates from 1.6 MW to 5.2 MW. The Fraser firs, which ranged from 9.7 kg to 12.6 kg, generated peak heat release rates from 3.2 MW to 4.3 MW.

Changing the orientation of the tree from vertical to horizontal, slowed the flame spread rate and reduced the peak heat release rate by more than 50%.

The ignition resistance of the trees with needles that had a percent moisture content in excess of 100% was demonstrated. No flames were spread beyond the area of ignition and the area of ignition self-extinguished. This was also consistent with the findings of the Scotch pine study.

The last objective of this study was to examine the impact of a residential sprinkler on the heat release rate of a dry burning tree in a compartment. A residential sprinkler with one of the lowest listed flow rates was chosen for use in this study.

Figure 19 shows a comparison of the heat release rates measured from the two room experiments. The sprinkler controlled the fire, prevented flashover and limited the heat release rate to less than 1.8 MW. Other fuels in the room were prevented from igniting. The non-sprinklered experiment had a peak heat release rate of more than 6 MW, which was reached in approximately 70 seconds after ignition. All of the fuels in the room were involved in the fire.

It should be noted that only a single sprinkler was used in this experiment. In a typical installation, more than one sprinkler would be exposed to the ceiling jet created by the rapidly developing dry tree fire. Hence it would be likely that more than one sprinkler would activate. However, in an actual installation of an NFPA 13D system, the flow rate of the system is designed to supply the listed flow rate for at least two sprinklers. Hence the first sprinkler to activate would have a larger flow rate than subsequent activations, potentially having a greater initial impact.

In 1982, dry tree experiments were conducted in two single story, ranch style model homes in Scottsdale, AZ [18]. In each case, a dry Christmas tree was ignited in the living room. In one of the experiments, two sprinklers activated and controlled the fire within 45 seconds. In the other experiment, 6 sprinklers were activated before the fire was controlled. Even though more than two sprinklers were activated, the fire was limited to the tree itself, similar to the results of the other Scottsdale test and the one room experiment presented in this report.

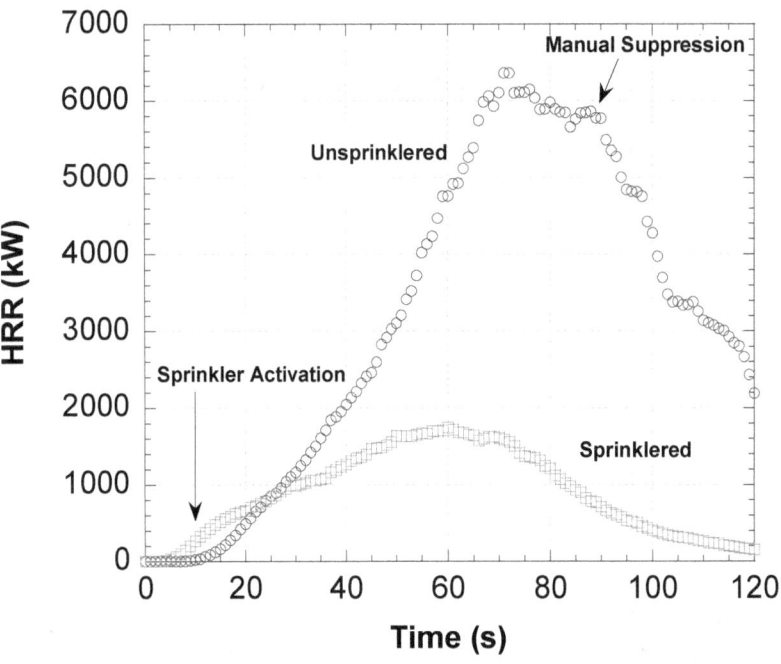

Figure 19. Sprinklered and non-sprinklered heat release rate comparison

SUMMARY

Properly maintaining a cut Christmas tree is important to retaining a high moisture content in the needles of the tree to limit accidental ignition and prevent rapid flame spread. A tree which has dry needles can readily ignite with a flaming source and generate heat release rates that are capable of causing flashover in residential scale rooms.

The data from the sprinklered room experiment demonstrated that even under conditions of extreme fire growth, approximately 1 MW within 10 seconds after ignition, a single sprinkler was able to prevent flashover, control the tree fire and limit the spread of fire to other objects.

ACKNOWLEDGEMENTS

The author would like to thank Matt Bundy, Steve Kerber, Roy McLane, Jay McElroy, Laurean DeLauter, and Jack Lee for assisting with the conduct of the full scale experiments.

The support of U.S. Fire Administration, especially project officer Meredith Lawler is greatly appreciated. The cooperation of Ellis Schmidt from the National Christmas Tree Association and Thomas Deegan from Viking Sprinkler Corporation proved useful in the conduct of these experiments.

REFERENCES

1. Ahrens, M., Home Christmas Tree and Holiday Light Fires. National Fire Protection Association, 1 Batterymarch Park, Quincy, MA, November 2007.
2. Hinesley, E. and Chastagner, G., Christmas Trees. Agriculture Handbook Number 66, Gross, K.C., Wang, C.Y., and Saltveit, M., eds. U.S. Department of Agriculture, updated May 5, 2004. http://www.ba.ars.usda.gov/hb66/title.html. Downloaded November 24, 2006.
3. Chastagner, G., *Factors Affecting the Moisture Levels of Cut Christmas Trees* as referenced in Ignition Handbook, Vytenis Babrauskas, Fire Science Publishers, Issaquah, WA, 2003 pp 838-839.
4. White, R.H., DeMars, D., and Bishop, M., Flammability of Christmas Trees and Other Vegetation. Proceedings of the International Conference on Fire Safety, Vol.24, Columbus, OH., July 1997.
5. National Christmas Tree Association, "How to Care for Your Farm-Grown Fresh Christmas Tree". http://www.christmastree.org/care.cfm. Downloaded, November 24, 2006.

6. Stroup, D.W., DeLauter, L., Lee J., and Roadarmel, G. Scotch Pine Christmas Tree Fire Tests, Report of Test FR 4010. National Institute of Standards and Technology, Gaithersburg, MD December 1, 1999.
7. Damant, G. and Nurbakhsh, S., "Christmas Trees – What Happens When They Ignite?". *Fire and Materials*, 18 pp 9-16, 1994.
8. WeatherUnderground, http://www.wunderground.com/weatherstation/wx dailyhistory.asp?ID=KMDMONTG1, Downloaded November 11, 2007.
9. Computrac® Max® 200XL Moisture Analyzer User's Manual. Arizona Instrument LLC, Tempe, AZ., November 2002.
10. Model RC-1E Owners Manual, Delmhorst Instrument Company, Towaco, NJ., March 2002.
11. Bryant, R.A., Ohlemiller, T.J., Johnsson, E.L., Hamins, A.H., Grove, B.S., Guthrie, W.F., Maranghides, A., and Mulholland, G.W., *The NIST 3 MW Quantitative Heat Release Rate Facility: Description and Procedures*, National Institute of Standards and Technology, Gaithersburg, MD, NISTIR 7052, September 2004.
12. Pitts, William, M., Annageri V. Murthy, John L. de Ris, Jean-Rémy Filtz, , Kjell Nygard, Debbie Smith, and Ingrid Wetterlund. *Round robin study of total flux gauge calibration at fire laboratories,* Fire Safety Journal 41, 2006, pp 459-475.
13. National Type Evaluation Program, Certificate Number 00-075A1, Model K-series, Mettler-Toledo, Worthington, OH., December 27, 2002
14. Omega Engineering Inc., *The Temperature Handbook*, Vol. MM, pages Z-39-40, Stamford, CT., 2004.
15. Blevins, L.G., "Behavior of Bare and Aspirated Thermocouples in Compartment Fires", *National Heat Transfer Conference, 33rd Proceedings*. HTD99-280. August 15-17, 1999, Albuquerque, NM, 1999.
16. Pitts, W.M., E. Braun, R.D. Peacock, H.E. Mitler, E. L. Johnsson, P.A. Reneke, and L.G.Blevins, "Temperature Uncertainties for Bare-Bead and Aspirated Thermocouple Measurements in Fire Environments," *Thermal Measurements: The Foundation of Fire Standards. American Society for Testing and Materials (ASTM). Proceedings.* ASTM STP 1427. December 3, 2001, Dallas, TX.
17. Viking Technical Data Form No. F_082197, Sprinkler 150 a-b. The Viking Corporation, Hastings, MI., March 19, 2004.
18. Ford, J., "Automatic Sprinklers: A 10 Year Study". Rural/Metro Fire Department, Scottsdale, AZ, 1997.

www.ingramcontent.com/pod-product-compliance
Lightning Source LLC
Chambersburg PA
CBHW081820170526
45167CB00008B/3471